THE GREATEST ANSWERING MACHINE MESSAGES OF ALL TIME

Written & Compiled
by
JOHN CARFI
&
CLIFF CARLE

ILLUSTRATED BY DON SMITH

CCC PUBLICATIONS • LOS ANGELES

Published by
CCC Publications
1111 Rancho Conejo Blvd.
Suites 411 & 412
New Bury Park, CA 91320

Manufactured in the United States of America

Cover ©1993 CCC Publications

Interior illustrations © 1993 CCC Publications

Cover & interior art by Don Smith

Interior layout & production by Oasis Graphics

ISBN: 0-918259-54-1

If your local U.S. bookstore is out of stock, copies of this book may be obtained by mailing check or money order for $4.95 per book (plus $2.50 to cover postage and handling to: CCC Publications; 1111 Rancho Conejo Blvd., Suites 411 & 412; New Bury Park, CA 91320.

Pre-publication Edition - 9/93

Fourth Printing - 3/96

DEDICATION

*This book is dedicated
to our mothers,*
ELINOR CARLE
&
YOLANDA CARFI,
*who made this
all possible
by making us
possible.*

INTRODUCTION

It was over ten years ago that we wrote our first book of Answering Machine Messages. Back then, in our wildest dreams, we could not have imagined how popular these machines would become. Today, many homes have more than one. There's the parents' machine, one for each of the kids — some families even have a machine for their pet...

"Scruffy isn't in his doghouse right now, he's out chasing cars, but if you have a bone to pick with him, bark a message at the tone..."

After three successful books and seven cassettes of funny messages for machines — plus another book on "getting even" for those who still can't accept the machine's existence — we hung up our guns, so to speak. Yessir, after about a thousand messages, we felt the world had had enough. That is, until our publisher $ugge$ted that we write and compile an anthology of the greatest messages. Hence the book you are holding in your hands.

We hope you enjoy recording these messages! We hope you and your callers share a buzillion laughs! And most of all, as was our original intent, we hope you have **no hang-ups!**

— JOHN CARFI & CLIFF CARLE

— CONTENTS —

HOW TO USE THIS BOOK

• Messages within chapters are conveniently categorized with headings and in alphabetical order to help you quickly find the right situation or mood.

• Always tailor the message to fit YOU. On first reading, a message may not seem to work for you or your situation. But, in most cases, you will be able to adapt them with only a slight change in wording.

For example, a message originally written for a male can usually be turned around to suit a female by slightly rewording it (e.g. switching a phrase such as "my girlfriend" to "my boyfriend.")

• Most of the messages will require no special talent. Just press the "Record Outgoing Message" button on your machine and start reading.

• Other messages [AS INDICATED] will require a little extra talent (SPECIAL VOICE or INFLECTION) or a little extra effort {using a PROP}, but it'll be worth it — you will definitely leave your callers thinking, "Hey, that's a GREAT message!"

FAMOUS MESSAGES
(Make Your Machine A Star)

ARNOLD SCHWARZENEGGER MACHINE #1

(AUSTRIAN ACCENT)

The machine you have reached is a Terminator X-3. I am right now computing your vital statistics and entering them into my memory bank. I now know who you are and where you live. So leave a message or... **I'll be back!**

BEEP...

ARNOLD SCHWARZENEGGER MACHINE #2

(AUSTRIAN ACCENT)

This is a powerful **Arnold Schwarzenegger** Answering Machine. If you don't leave a message your phone service will be TERMINATED!

Hosta la-vista, baby!

BEEP...

[NOTE: Arnold's so big he gets **two** messages]

CLINTON, BILL

{PROP: TYPEWRITER}

[TYPEWRITER CLICKING IN B.G.]

Hello. This is the Answering Machine Wire Service. If you have anything newsworthy to report, do it after the beep — but first, this just in:

Clinton, today, proposed another Gay Rights Bill. As expected it was vetoed by the House. Congress, meanwhile, is working on a bill for **Bisexual** Rights — it could go either way...

BEEP...

DOCTOR RUTH MACHINE

(HIGH-PITCHED VOICE — EUROPEAN ACCENT)

This is the **Dr. Ruth** Model Anserine Machine. Just look how you're holding the phone! It shows definite signs of latent sexual timidity. C'mon! Don't be afraid! Put the receiver close to your mouth! Closer! It's not going to hurt you! Now open your mouth very wide! Wider! That's good! Now say something raunchy, 'cuz I'm a horny ol' machine who likes to have good phone sex!

BEEP...

DON RICKLES MACHINE

(CRASS)

You've reached a **Don Rickles** Model Answering Machine. What's-a-matter hockeypuck, you can't leave a message? Great! I got someone calling with the I.Q. of a plant! What a knucklehead! I suppose you need to be spoonfed, too? You know what a **beep** is... when you hear one, clown, leave a message!

BEEP...

DRACULA

(BELA LUGOSI VOICE)

Gooood evening. This is Count Dracula. These recorded messages drive me batty! So I stepped out for a bite. If you don't mind sticking your neck out, leave your name and number and you can be bloody sure I'll get back to you... before sunrise... because I have a stake in this machine... don't cross me... Ah-ha-ha-ha-ha!

BEEP...

GODFATHER MACHINE

[OPTIONAL: VIOLIN MUSIC IN B.G.]

(SCRATCHY VOICE)

This is the **Godfather** Answering Machine. Nobody's in right now. They're out taking care of someone who was disrespectful — he hung up on me. Now, I hope you's gonna show me some respect... or you get to go deep sea diving with Jimmy Hoffa... capice?

BEEP...

JOAN RIVERS MACHINE

(FAST)

Hi. This is a **Joan Rivers** Model Answering Machine. Can we talk? I looked up the word **ugly** in the dictionary and it had your picture beside it. Oh! Oh! Can we talk? You're so fat, they're gonna name a planet after you! Oh! Oh! can we talk? You're so stupid, you think **foreplay** is sex with three other people! Oh! Oh!

(TO SELF)

Gee, I wonder why nobody ever leaves a message???

(TO CALLER)

Can YOU talk?

BEEP...

LIEUTENANT COLOMBO MACHINE

{PROP: SHEET OF PAPER}

[OPTIONAL: "MYSTERY" MUSIC IN B.G.]

(OBSEQUIOUS COLOMBO ATTITUDE)

Hi. this is the **Lieutenant Colombo** Model Answering Machine. I'm having a little problem here — maybe you can help me? Now, you're calling me, so you're gonna leave a message, right? Okay. But there's one thing that **bothers** me... let's see...

[SHUFFLE PAPER]

I got it here somewhere... Ah!

Last week you didn't leave a message. This makes me **very** suspicious. I mean, if you're not guilty, you should have no problem leaving a message at the tone, right?

[SHUFFLE PAPER]

Oh, just one more thing...

BEEP...

MADONNA MACHINE

[MADONNA SONG "LIKE A VIRGIN"
PLAYING IN B.G.]

(SEXY FEMININE VOICE)

Hi there. This is the **Madonna** Model Answering Machine. If you leave your name and number and don't do anything naughty, like hanging up, I might let you come over some-time and play with my knobs... for the very first time.

BEEP...

MAE WEST MACHINE

[SLEAZY "BEDROOM" MUSIC IN B.G.]

(MAY WEST ATTITUDE)

Ooooo, this is a **Mae West** Model Answering Machine. Why don'cha call me up 'n' speak to me sometime, big boy?

You know, I'm a little shy — so on a first date I just like to hold each other's **glands!**

Ooooo, why don'cha open wide and show me what-cha got... on your mind, big boy!

BEEP...

NO TELLING WHERE
I'M HANGIN' OUT!
BUT I'M NOT HERE.
MAYBE I'M HAVING
A PERRIER

PEROT MACHINE

{PROP: DOOR IN B.G.}

(TEXAS DRAWL)

Howdy! This is a **Ross Perot** Machine. NAME is not in...

[OPEN DOOR]

Wait — Yes! he **is** in...

[CLOSE DOOR]

No! — he went back **out**

[OPEN DOOR]

Uh... Yes! Now he's coming back **in**... but wait, he's hesitating and...

[CLOSE DOOR]

No! He's backing **out,** again!

(EXASPERATED)

Listen here, I can't take any more of this pussyfootin'! Leave a message and I'll have him ring y'all when he finally decides whether he's **in** or **out!**

BEEP...

REVEREND JIM

("REVEREND JIM" - FROM "TAXI" - VOICE)

Uh, Alex! Alex! You know what, Alex? It's happening again... **the voices.** I'm hearing **the voices** again. But here's the strange part, Alex: just before I hear **the voices,** there's always this weird beep...

BEEP...

RICH & FAMOUS MESSAGE

(ROBIN LEACH VOICE)

This is Robin Leach, and welcome to "Phone Calls of the Rich and Famous." This finely appointed answering machine is prepared to record your champagne wishes and caviar dreams. Oh listen! I think I hear a **famous** caller now! Why yes, it's...

BEEP...

ROCK STAR MESSAGE

[PLAY "LIVE VERSION" OF ROCK GROUP WITH FANS CHEERING IN B.G.]

(EXCITED)

Hey, thanks for the call. I can't pick up the phone 'cuz I'm ON in ten seconds — and I've got 50,000 screaming fans waiting...

[FADE OUT MUSIC & CHEERING]

(NORMAL VOICE)

Okay, so I got a desk job. But I can dream, can't I?

BEEP...

RODNEY DANGERFIELD MACHINE

(FAST AND NERVOUS)

Hi. This is **Rodney** the Answering Machine. I tell ya, It's tough being me! The other day my owner says, "I got good news and bad news. The **good** news is, I'm replacing you with a newer model."

I said, what's the **bad** news? He said, "my loudmouth **mother-in-law** needs a machine!" Sheez! that's my trouble — I never get any respect!

BEEP...

I DON'T GET
NO RESPECT!
NO RESPECT!
MY WIFE
CALLS ME UP
JUST TA HANG UP
ON ME.

ROD SERLING

(DEEP, MESMERIZING VOICE)

You're dazed, bewildered, trapped in a world without time, where sound collides with color and shadows explode. You see a signpost up ahead — this is no ordinary telephone answering device... you have reached "The Twilight Phone."

BEEP...

ROSANNADANA MACHINE

(NASAL VOICE)

Hi. This is a **Rosannadana** Model Machine. Are you calling from a payphone? I hope you're not 'cuz maybe the person before you needed change — so they put the receiver under their armpit while they searched for the coins. Then you come along and the receiver stinks like B.O. Or sometimes you look at the mouthpiece and there's all these little itty-bitty pieces of crusty food in the little holes and it just makes you wanna die!

BEEP...

ROSEANNE ARNOLD MACHINE

(WHINY VOICE)

This is a **Roseanne** Model Machine — why am I answering the phone — it's not my job — let that lazy husband of mine get it! Hey! Where is he? And where are those good-for-nothin' kids? They're suppose to be cleaning up this dump! I ain't got time for this crap! I gotta go down to the beauty parlor an' get an estimate!

BEEP...

60 SECONDS

{PROP: OVENTIMER OR LOUD TICKING CLOCK}

[START CLOCK TICKING]

Welcome to "60 Seconds," the answering machine news program, with Mike Wallet and Sam Donalddaughter. Unfortunately, today we only have time for Andy Fooey:

[TURN OFF CLOCK]

(WHINY VOICE)

Don't you hate it when you want to talk to a friend and you reach a machine? Why do people **have** these machines? I hate machines! And don't you hate people who have real long outgoing messages? I hate long messages! And did you ever notice that..

BEEP...

SOAP OPERA MESSAGE

(TV ANNOUNCER VOICE)

Hello. Welcome to "Phone Calls Of Our Lives." Today the life of the <u>NAME's</u> and today's burning question: "Will they call back the strange people who call them? Or will they callously leave their friends on hold, as usual?" For the answer to these and other important questions, leave a message at the tone.

BEEP...

STAR TREK

(CAPTAIN KIRK VOICE)

Tape... the final frontier. These are the messages of <u>NAME's</u> machine, its five second mission, to seek out your name and number. To boldly record what no machine has recorded before!

BEEP...

TV OVERDOSE

Sorry, I can't pick up the phone. I'm disciplining **All My Children.** I don't want them to take after me and watch too much TV. I want them to grow up living **Lifestyles Of The Rich And Famous.** I mean **Gimme A Break,** they only have **One Life To Live** — and remember, **Father Knows Best!**

BEEP...

TRICKY MESSAGES
(For Those Who Can <u>Give</u> A Joke)

ABSENT-MINDED MESSAGE

{PROP: DESK DRAWERS}

Hi. This is <u>NAME</u>. Please leave a message at the...

(PAUSE – FRUSTRATED)

Now where'd I put that **beep**???

[OPEN AND CLOSE 2 DRAWERS OVER:]

Not there...

Not there...

(RELIEVED)

Ah! Here it is...

BEEP...

ALIEN MACHINE

(WORRIED)

Listen, my machine has been acting very strange lately. It's almost as if... you're gonna think I'm crazy... as if **aliens** have taken...

(NEW MECHANICAL VOICE)

I'm sorry. Disregard the previous transmission. I'm encountering excessive stress today. At the electronic tone, log your name and the coordinates of your telephone unit...

BEEP...

ANALYTICAL MESSAGE

Hello, it's me. If **you're** who I think you are, leave a message. If you **aren't** who I think you are, leave a message anyway. If **I'm** not who you think I am, so what — leave a message. If **you** aren't who **you** think you are, don't leave a message — go get professional help.

BEEP...

BURGLAR

[FOR BEST EFFECT HAVE A FRIEND MAKING
MUFFLED CRIES OF "HELP" IN THE B.G.]

(THUGS'S VOICE)

<u>NAME's</u> residence. Dis is Lefty — I'm kinda like, ransackin' dis apartment, y'know? and NAME is kinda... **tied up** right now, heh, heh! So why don-cha leave yer name and number an' he'll probably call youse back when he, uh, gits **free**, heh, heh!

BEEP...

CHEAP MACHINE

(EXCITED)

Hi. This is my new machine! Boy did I get a great deal! You won't believe what I paid! Incredible! I don't see how that fool on the corner could possibly sell these machines for only...

(DRAW OUT WORDS AND MAKE "STATIC BREAK-UP" NOISES)

Teeeeeeeeeeeeeeeeeeeeeeeeeennnnnnnnnnnnnnnnnnnnnnnnnnnn buuuuuuuuuuuuuuuuuuuuccccccccccccccccccccccccccccks!

BEEP...

CLASSICAL MACHINE

(SNOBBISH)

Hello and welcome to Dial-A-Concerto. Today we will be enjoying a selection for unaccompanied answering machine: the renowned **"One Note** Concerto." As always, your comments are welcome immediately following the performance:

BEEP...

CLICHES

(SLIGHTLY IRRITATED)

<u>NAME</u> here. You know what really bugs me is people who use old hackneyed cliches in their messages and think they're clever. Boy I hate that! Cliches are a dime a dozen. If you've heard one, you've heard them all. Anyway, leave a message at the tone — and "Have a nice day!"

BEEP...

COMPUTER DATE

(MECHANICAL AND CHOPPY)

Hello. I'm a computer... So, uh, what's your sign? Call here often? You've probably met other computers, but wait'll ya get a load of **my** HARD-drive! So, uh, you wanna go out? Before you say NO, just answer me this... have you ever TRIED computer dating?

BEEP...

CONFUSE THE CALLER

[OPTIONAL: LIGHT CHAMBER MUSIC IN B.G.]

Hello. You are listening to **today's** message. If you want to hear **yesterday's** message, just call back tomorrow.

BEEP...

DOCTORED MESSAGE

[HAVE SOMEONE OF THE **OPPOSITE SEX** RECORD THIS MESSAGE FOR YOU]

(WORRIED)

Hi. This is <u>NAME</u>. Please leave a message and I'll call you in a few hours. I have to go back to my doctor's office. I think he gave me the wrong... **hormone shot!**

BEEP...

DRUNK MACHINE

(SLURRED VOICE)

(HICCUP) I'm starry, but the slumber you have breached is not a jerking wonder. (HICCUP) Please console your disectory, or cleave a message at the bone and clang hup. (HICCUP) Tank shoe.

BEEP...

GAME SHOW

[OPTIONAL GAME SHOW MUSIC IN B.G.]

(TV ANNOUNCER VOICE — FAST)

Hey! You've reached the "Answering Machine Game Show" where callers get a chance to win fabulous prizes! When you hear the tone, if you can answer this question correctly: "What is your name and number?" you will win a luxurious and exciting **phone call** from <u>NAME</u>! And now, **here's** our first contestant...

BEEP...

HAYWIRE MACHINE

{PROP: MISC. TOOLS}

[RATTLE TOOLS IN B.G.]

(SLIGHTLY ANGRY)

I hope you're getting this message 'cuz this crazy machine has been playing backwards! I've been trying to fix it all day and I think it's finally working so, **tone a message at the Leave... you Thank!**

BEEP...

HYPNOTIC MESSAGE

[PLAY EERIE MUSIC IN B.G.]

(DEEP VOICE)

Your eyelids are getting heavy... You're entering a d-e-e-e-p sleep... thinking of nothing but the sound of my voice. Tonight at midnight, you will get up and write me a $1000 check. You will mail the check to me in the morning, on your way to work. When you hear the beep, you will awake and remember nothing.

BEEP...

IDLE THREAT

Hi. I should warn you that I wired my answering machine to my VCR and you are being videotaped right now, so I'll know who you are. Don't believe me? Okay, I'll prove it: if you hang up... uh??? I won't call you back!

BEEP...

MISPLACED MESSAGE

(PERPLEXED)

You know, I had a really funny message for you today, but now I can't seem to find it...

[PAT YOURSELF DOWN]

Darn! I must have left it in my **other** machine!

BEEP...

PHONEY OPERATOR

(NASAL VOICE)

You have reached <u>TELEPHONE NUMBER</u>. If you have a rotary phone, the number once again is <u>TELEPHONE NUMBER</u>. Or, if you have "Touch Tone," the number is...

(SING)

"beep, beep, boop—bimp, berp, bapp, beep."

BEEP....

SHORT QUIZ

Hi. I hope you don't mind, but this is a Math-O-Matic machine. The way it works is, if you answer this short algebra quiz, you can leave a message. Ready? Okay, how much is 5Q plus 5Q?

[PAUSE FOR CALLER TO THINK: " 10Q "]

Your welcome!

BEEP...

SMART MACHINE

[DISGUISE YOUR VOICE - OR HAVE SOMEONE ELSE RECORD FOR YOU]

Hi. This is <u>NAME</u>'s answering machine. <u>NAME</u> isn't in right now, but if you leave your name and...

[PAUSE]

Oh, wait a minute — here she comes now!

[PAUSE]

No, sorry, I thought that was her.

Well, give me your name and number and I'll tell her you called. BEEP...

STRANGE VISITORS

(WHISPER)

You won't believe this, but I've just been visited by Aliens! Yeah, no kidding! As a matter of fact, they come every week — and clean my house. The leader's name is Maria. The other is Juan.

BEEP...

SURPRISE INSIDE

Hi. Leave a message and...

(SURPRISE)

wait a minute! What's this??? Oh, my God! I just opened that little tape lid on my brand new answering machine and guess what I found??? A **syringe!**

(TO SELF)

Where's the phone book???

[SHUFFLE PHONE BOOK PAGES]

(CALLING OUT TO PERSON IN B.G.)

Honey? How do you spell "Jacobee & Myers???"

BEEP...

TEASE CALLERS

(SEXY VOICE)

Hi. This is <u>NAME</u>. Congratulations! I gave you the **right number.** But now you only have 30 seconds to convince me why I should call you back... Good luck!

BEEP...

WIRE TAP

[TAP DANCING — AND/OR TAP DANCING MUSIC — IN B.G.]

Hello. I'm, uh, trying to maintain a **low profile** these days, so I won't be picking up my phone. Please leave a message and I'll call you back when I can. But be careful what you say — I hope you don't think I'm paranoid, but sometimes I get the feeling someone's tapping my phone.

BEEP...

HIP TEEN
(Radical Lines For Young Minds)

ATTENTION!

(SING)

"From the halls of Montezuma,

To the shores of Tripoli..."

(SPEAKING VOICE)

Like the Marines, I, too, am looking for **a few good men!** If you qualify, leave your name and number and I'll get back to you.

BEEP...

BRAIN CHEESE

(LAUGHING)

Hey dude, glad you called! I got a way cool joke for you, man! It's so funny, you'll laugh till yer brain turns to Velveeta!

　　[PAUSE]

Whoa, sorry dude, guess you already heard it!

BEEP...

BIRTHDAY SUGGESTIONS

("SURFER" VOICE)

Hey dude, it's like, my birthday and here's a list of what you can get me:

1. Alice-In-Chains T-shirt

2. Doc Martin boots

3. Any Heavy Metal CD

But, like, don't buy me a **book**, Dude — I already got one.

BEEP...

CHOICES

Hi. This is <u>NAME</u>. I can't pick up the phone right now because:

A. I'm not home.

B. I'm outside. Or,

C. I'm having sex.

Mom, if it's you calling, it's not C!

— but if it was C, it would be "safe C!"

BEEP...

HOT BABE

Hi. This is <u>NAME</u>, the sultry blonde with a bust that could cause wide-spread civil unrest, and thighs that could be the basis for a worldwide religion. I can't come to the phone right now as I'm either at the Mall spending all of Daddy's money or I'm admiring my fabulous bod in the mirror. Please leave your message at the "moan" and I just might get back to you. Oh, by the way, how did you get this number?

BEEP...

LATE SLEEPER

[LOUD, HEAVY METAL MUSIC IN B.G.]

Hi. I'll be sleepin' in this morning — I had a wild party last night! Man, talk about rude neighbors! The guy below me was poundin' on his ceiling all night! I hadda keep cranking up the stereo to drown him out!

BEEP...

LUSTY LIBRARIAN

Later man! I'm off to the Library — but it's not what you think — there's this new, knock-out, drop-dead Librarian. She's so sexy, when she reads, she moves her **hips!**

BEEP...

NO LIFE

(SOLICITOUS)

Hello. I am home right now, and could easily pick up, but leave your message after the beep anyway — my TV's broke and this machine is my only form of entertainment.

BEEP...

NOVEL MESSAGE

(EXCITED)

Hi. You'll have to leave a message. I can't come to the phone because I'm totally engrossed in this new, exciting **novel** I just picked up. It's written by some guy named **Webster**, and get this: all the words are in **alphabetical order!**

BEEP...

OUT TO LUNCH

Hey man, you missed me — I'm out to lunch... nah, I don't mean permanently! I'm kinda into a health and nutrition mode, today —so I'm eatin' something from one of the four basic food groups: McDonalds, Burger King, Jack-In-The-Box, and Taco Bell.

BEEP...

PARTY ON!

{PROP: DRINKING GLASSES}

[LOUD STEREO IN B.G. - HIP, PARTY MUSIC]

[CLINK GLASSES TOGETHER NEXT TO YOUR MACHINE'S MIKE]

(TO CALLER)

Oh, hi — excuse me a minute...

(TO IMAGINARY GUESTS IN B.G.)

Sorry Babes/Guys, party's over. You'll have to put your clothes back on and leave so I can clean up.

(TO CALLER)

Still there? Boy did you miss a wild one tonight! But leave your name and number and I'll make sure you're invited to my **next** killer party!

BEEP...

PICK ONE

Hi. I'm not in. I'm down at the corner phonebooth making prank calls. Here's one you can try: look up the name **Booger** in the phonebook and call. If the guy's name is **Bob,** ask for **Mike.** The guy will say, "There's no one here named 'Mike.'" You say, "Sorry, I must have **picked** the wrong **Booger!**"

BEEP...

QUICK STUDY

Hey Dude, I'm off to Blockbuster. My English Comp. teacher said I gotta buy a dictionary for class — but I'm gonna go check if it's out on video yet.

BEEP...

RAP MESSAGE #1

[TO A RAP BEAT]

Please have pity on my poor virgin machine,

Who gets messages that are mostly obscene.

If you speak at the beep,

Nothing sordid or cheap,

I'll return all the calls that are clean!

BEEP...

RAP MESSAGE #2

[TO A RAP BEAT]

The human is gone,

You've reached a machine.

Here comes the beep,

You know the routine...

BEEP...

SMART MESSAGE

Stay calm and don't panic. In about 20 seconds there will be a short oral exam. Here's a few quick-study tips: start with your name, number and reason for calling. However, since you obviously have already paid the 25 cent exam fee, feel free to flunk by hanging up — but, then you'll never find out if you really passed!

BEEP...

SMELLY MESSAGE

Hi. I'm not here right now, but if you don't feel like **talking** to a machine, don't worry — **mine's** hooked up to the Rimco "Gas-O-Matic" Decoder. Just "let one loose" into your receiver and the decoder will **analyze** whose it is. But if you ate beans last night, PLEASE, just leave your name and number!

BEEP...

ROOMIES #1

(FAST)

Sorry you missed us. The two of us are so busy we're often coming and going. And unfortunately, the one going has already gone and the one coming ain't here yet.

BEEP...

ROOMIES #2

A. JANE, I can name that tune in **three** notes!

B. JOHN, I can name that tune in **two** notes!

A. JANE, I can name that tune in **one** note!

B. Okay, JOHN, here's your **one** note:

BEEP...

ROOMIES #3

A. Okay, <u>JOHN</u>, it's your turn to record the message.

B. I'm not in the mood, <u>JANE</u>, you do it today.

A. I do it **every** day — you always put it off — you're such a procrastinator!

B. A what?

A. Procrastinator! Don't you know what **procrastination** is?

B. Uh, I was gonna look it up once, but I never got around to it. Maybe tomorrow???

BEEP...

ROOMIES #4

(FIRST ROOMMATE)

Hi. This is <u>JOHN</u>...

(SECOND ROOMMATE)

And this is <u>JANE</u>...

(FIRST ROOMMATE)

today, we're coming to you in all new...

(BOTH)

"Answering Machine Stereo!"

We're both taking off right now, however, **one** of us will return your call later...

(SECOND ROOMMATE)

but it'll probably be in **mono.**

BEEP...

MESSAGES FOR GOOD SPORTS

(You'll Have A Ball!)

AEROBIC MACHINE

[AEROBIC MUSIC IN B.G.]

(FAST)

Hi! Welcome to the Aerobic Answering Machine Workout Tape. Let's start by doing a few curls with your receiver...

And 1, and 2, and 3, and 4!

Alright! You're looking good! Now it's time to work those jaw muscles up and down...

And 1, and 2, and 3, and 4!

C'mon, faster! Let's get rid of those fat lips!

Now, on the beep I want you to shout out that message!

C'mon, **louder,** I can't hear you!

BEEP...

BLOW BY BLOW

[ALL FAMILY MEMBERS ARGUING IN B.G. — ONE FAMILY MEMBER SPEAKS CLOSE TO MIKE]

(FIGHT ANNOUNCER'S VOICE)

It's a right to the jaw... a left jab to the head... it's getting ugly folks! Whoa, there goes a vicious uppercut... There's blood everywhere... I can't believe the referee is letting this go on!

(NORMAL VOICE)

Oh, hi! This is the LAST NAME's. We're in the middle of one of our warm, nurturing family discussions — so please leave a message at the **bell** — I mean the **beep.**

BEEP...

BEAT THE BUZZER

(FAST, PROFESSIONAL SPORTS ANNOUNCER VOICE)

Well sports fans, it doesn't get any better than this! There's 10 seconds left to go in the tape and NAME has possession of the mike! The question on everyone's mind: will he get his outgoing message off before the final buzzer? NAME sees an opening — he clears his throat — a little head fake — he opens his mouth to shoot off a remark and... 3... 2... 1 — Oh no! His sinus is blocked!

BEEP...

DRIVING YOU CRAZY

Hi. I'm out shopping and my husband's over at a friend's watching car racing. What a stupid sport! Take the Daytona **500** — why don't they just make it **shorter** so they don't have to drive so fast?

BEEP...

FIRST GAME

This is <u>NAME</u>. Last week my boyfriend took me to my first football game. What a ridiculous sport! All those guys hitting, kicking, shoving, tearing uniforms and fighting over a pigskin! I mean wouldn't it be easier if they just gave each team their own ball?

BEEP...

GREAT OUTDOORS

Sorry I'm not in, I went over to Echo Canyon... Sorry I'm not in, I went over to Echo Canyon...

BEEP...

GUILT TRIP

Hi pal, I'm out white-water rafting, but if you're a **real** friend, while I'm gone, you'll help me with a list of things I need done around the house. You only have to do **two** of them:

wash car

paint fence

clean gutters

leave name... and,

leave message.

Your choice, **Bud?** pick two!

BEEP...

HALF-TIME REFRESHMENTS

(FAST)

Hi. You'll have to leave a message, we're running late! My wife finally talked me into going to this stupid **ballet** thing with her. Dang, I didn't have time to eat, so I guess I'll have to grab a hotdog and beer when we get there.

BEEP...

HIGH SCORE

You know, a lot of snooty people say bowling is a boring sport, but you wouldn't believe what happened at the bowling alley last night...

As I was changing shoes, this sexy Blonde, looks just like Diane **Lane,** walks up. To my shock, she tries to **pin** a paternity rap on me. I knew I was being **framed!** I asked her what **gutter** she crawled out of and turned to make a **break** for the back **alley.** Suddenly, I **split** the crotch of my pants. I figured that was another **strike** against me. But luckily for me, I had a **spare.** As I made a quick change, she saw the **score** and quickly changed her mind. We **rolled** back to my place and **ball**ed all night.

BEEP...

HOME RUN

(MALE, SINGING)

"Take me out to the ball game, take me out to...

(MALE, NORMAL VOICE)

Oh, hi! Well, it's Summer and you know what that means...

"America's Favorite Pastime!"

(SEXY FEMALE VOICE)

Okay Honey, I'm ready to **play,** now!

(MALE, CONFIDENTIAL VOICE)

Well, looks like I'm up to bat — hope I get to touch all the bases!

BEEP...

JOHN MCENROE MACHINE

This is a **McEnroe** machine and I just wanted to let you know that...

(SHOUTING)

<u>NAME</u>'s not **IN!**

Are you blind?

He's clearly **OUT!**

What a stupid call!

Leave a message, you jerk, then go to...

BEEP..

MANLY MAN

What's happenin'? Bet you expected to catch me sittin' at home watchin' TV. Well you got the wrong guy, I'm a man of action! I like to live on the edge. Matter of fact, as you're listening to this message, I'm behind the wheel of a Formula I race car. But I should be home pretty soon — the dang thing is suckin' up all my quarters!

BEEP...

MIND ON THE GAME

[TV TUNED TO FOOTBALL GAME IN B.G.]

Hello. It's <u>NAME</u>. I'm out and my husband won't be picking up 'cuz he's got a game on TV. You know, I think he's been watching a little **too much** football lately. Last night I told him I saw the New York Philharmonic play Beethoven. He asked me, "Who won?"

BEEP...

PROBLEMS SCORING

Hi. I'm out bowling — a sport that makes **sense!** Last week, three of my friends finally talked me into playing golf. I've never seen such a bunch of sore losers! They tried to tell me **I lost,** when I had the **highest** score!

BEEP...

RUNNING GAG

Hi. <u>NAME</u> here. It's such a beautiful day, I'm out by the pool, sipping a tall cool one. A couple friends tried to talk me into running with them. But I hate jogging — the ice always flies out of my drink — and my cigarette keeps going out!

BEEP...

SOCCER SENSE

Hi. I'm out playing <u>FAVORITE SPORT</u>. Yesterday, I went to my first Soccer game. You know, I figured out how those foreigners invented it. They **stole** ideas from a bunch of American sports and made it into **one** game. Check it out: Soccer is 22 guys wearing **tennis** outfits and **track** shoes, kicking a **volleyball** on a **football** field into an **ice hockey** goal, but instead of **baseball** bats, they use their heads!

BEEP...

SPORTS QUIZ

Hi. This is <u>NAME</u> with a little sports quiz:

How many **miles** is the Indy **500?**

What **sport** do you use your **foot** to punt a **ball?**

And finally, How many **seconds** does a boxer have to get up during a **10**-count?

If you had to **think** about any of the answers — you must be calling my wife.

BEEP...

TEAM SPIRIT

Well, it's football season, so you'll probably have trouble reaching me in person. Today I'm watching the Tigers vs. the Bulldogs. Man, I envy those rugged sounding teams. I went to a small technical/science school — our mascot was Bacteria.

BEEP...

WORK OUT

Hi. I'm out getting some exercise. I'm at my peak now, but I had to work up to it. First, I started out walking. Then graduated to jogging, then to running. Now, I'm right where I want to be — I'm driving!

> [IF POSSIBLE, ADD HONKING HORN AT THE END OF THIS MESSAGE]

BEEP...

YOU BET

Hi. This is <u>NAME</u>. Boy, I gotta watch my drinking! Yesterday, a buddy and I were downing a few beers during a football game on TV and I lost a 50 dollar bet on a play. I guess I didn't realize how drunk I was, 'cuz a few seconds later, I lost another $50 —on the **instant replay!**

BEEP...

PHONEY SEX
(Cajole, Captivate, And Caress)

BRAND NEW MACHINE

Hello. I just bought a new answering machine and if you don't hang up, **yours** will be my first message. Now, if you finish your message too soon, don't worry — it happens to everyone now and then. And remember: the size of your message isn't important — it's your performance that counts.

BEEP...

BUSY BODY

(SEXY FEMININE VOICE)

Hi. This is Desiree. NAME can't come to the phone right now. He's, uh, taking care of some business...

(SEXY GIGGLE)

But if you leave your name and number, I'm sure he'll...

(SEXY MOAN)

take care of you too.

BEEP...

CENSORED

(TV ANNOUNCER VOICE)

The message you are about to hear is rated "R" — parental discretion is advised.

(SEXY VOICE)

Hi Sexy! I'd just love to **BLIP** your **BLIP** with my sensuous, pulsating **BLIP.** So leave a **BLIP** at the tone.

(TV ANNOUNCER VOICE)

This message was edited for telephone.

BEEP...

FIRST TIME

(GIRLISH VOICE)

Hi. Guess what? You're my very first caller. Yes, it's true, I'm a **virgin** machine. And you're so handsome, so strong, so assertive! I always dreamed my **first message** would be like this! One last thing, please be gentle???

BEEP...

FOR SALE

Hello. This is <u>INITIALS</u>. If you're a friend, at the tone, leave your name and your number. If you're calling about my ad, leave your measurements and your sexual preference.

BEEP...

HOME ALONE?

[HAVE FRIEND OF OPPOSITE SEX DO HEAVY BREATHING OR MOANING IN B.G.]

Hi. <u>NAME</u> here. I can't come to the phone. I'm having a terrific sexual experience and it's absolutely fantastic! I can't even imagine what it's going to be like when my girlfriend gets here!

BEEP...

HOT & STEAMY

{PROP: TEAPOT}

[START WITH TEAPOT SOFTLY WHISTLING UNDER LOW FLAME]

(SENSUOUS)

I'm getting hot... hotter...

[TURN UP FLAME TO INCREASE WHISTLING]

Oh yeah, I'm ready!

Hi. This is the teapot. I'm taking over for the answering machine who has the day off. Please leave a **steamy** message — 'cuz I'm really hot for you babe!

BEEP...

INNER SPACE

(MECHANICAL "ALIEN" VOICE)

Greetings Earthling. I am a being from outer space. I have trans-mutated into an answering machine. Right now I am having sex with your ear. I can tell you are enjoying it because you are smiling...

BEEP...

MAGIC GENIE

{PROP: BOWL OR POT}

[TALK INTO A BOWL OR POT FOR ECHO EFFECT]

(MONOTONE)

Hello. This is a Magic Answering Machine and I am the Genie inside. Your wish will be my command. Just rub your receiver three times — and when you hear the tone — leave your name, number and a wish. Ready? Okay, start rubbing... Yes, that's it... Okay, a little to the left...

(INCREASING EXCITEMENT)

Now harder. Now faster... Yes, faster, faster! Faster! Oh! Oh!

OHHHHHHHHHHHHHHHHHHHHHHHHHHHHHHH!

BEEP...

OBSCENE CALLER

(NORMAL VOICE)

Hi. If this is an obscene phone call, I expect your looking to elicit certain emotional responses from me to satisfy your twisted mind. Even though I'm out at the moment, I'll do my best:

(EXAGGERATED INDIGNATION)

Who is this??? That's disgusting!!! Hey, I don't have to listen to this!!! You ought to be put away!!!

(PAUSE - THEN, EXCITED)

You can do **what?** Listen, please leave your name and number — and whatever you do, don't leave your phone booth. I'll call you right back!

BEEP...

ORAL MESSAGE

(SUGGESTIVE ATTITUDE)

You know, although I like to do it up and down, my boyfriend prefers a back and forth motion. But either way we both achieve maximum stimulation and are left with a wonderfully ALIVE feeling. So leave a message and we'll return your call as soon as we've finished...

brushing our teeth.

BEEP...

PHONE SEX (LITERALLY!)

This is NAME's machine. She's out, as usual. I'm glad you called because I'm lonely and I'd like to be intimate with someone, okay sweetie? Good! On your "touch tone" phone, gently rub the #1... That's good! Now rub #3... Back to 1... Oh yeah, that's great! Now work your finger down to the "O" ... Yes, that's nice! More! More! Oh baby! Oh! Oh! Oh!...

(CALM VOICE)

So, after all we've just been through, don't you think I should know your name?

BEEP...

RIGHT NUMBER, WRONG PHONE

(SENSUOUS)

Ooooooo! you finally called! I get so... HOT when I hear your voice! I want your sexy body! Come closer, closer! Now, at the tone, tell me your wildest fantasy!

(NORMAL VOICE)

Uh, my apologies — this machine once belonged to a "900" Phone Sex service.

BEEP...

SAFETY TIP

(CASUAL)

Hi, leave a message... but first, a tip from the American Safety Counsel:

(PROFESSIONAL SOUNDING VOICE)

Remember to always **practice safe sex!** ... **Practice** at least three times a day — and to be sure you're **safe**, put an airbag on your headboard! Thank you.

BEEP...

SEXUAL OVERDRIVE

(BRAGGING)

Hi. It's <u>NAME</u>. You won't be able to reach me all weekend, because I have a tremendous sex drive...

(EXASPERATED)

my girlfriend lives 200 miles away!

BEEP...

SMART MACHINE

(MECHANICAL VOICE)

Hello, you have reached an Intele-matic Machine. I use a tonal code to inform you as to why the <u>NAME</u>'s are not answering their phone: **Three beeps** means they're outside and can't hear the ring. **Two beeps:** they're in the middle of a meal and want to finish. **One beep:** they're making love & don't want to be disturbed...

BEEP...

STUD (OR HOT BABE) MESSAGE

{NOTE: MAY REQUIRE TWO PEOPLE}

[LOUD, PERSISTENT KNOCKING ON DOOR IN B.G.]

Hi. I can't answer the phone because there's two gorgeous babes/hunks at my door...

[KNOCKING GETS LOUDER AND MORE FRANTIC]

one's French, the other's Swedish and they both can't wait!

[KNOCKING GETS EVEN LOUDER]

(DEJECTED)

So I guess I better go and let them **out.**

BEEP...

STUDYING HARD

Sorry you missed me, I'm at night school. I just signed up for a Sex Ed. class. I bet it's going to be real cool! — the brochure said the final exam would be **oral!**

BEEP...

RELIGIOUS REVELRY
(Don't Worry, You Won't Go To You-Know-Where)

APATHETIC MESSAGE

(SLOW, NONCHALANT)

You have reached the **Apathy** Society. Either leave a message or... hang up. It's all the same to us. You're probably wondering, "What is the Apathy Society?" We don't know — and we don't care.

BEEP...

CATHOLIC MACHINE

(IRISH ACCENT)

Hello, me fine caller. You've reached Dial-A-Confession. At the tone, be so kind as to leave your name, number and a brief sin...

BEEP...

ETHNIC MESSAGE

This is <u>NAME's</u> machine. Listen, don't hang up 'cuz if you're **Jewish**, you'll probably feel guilty afterwards — if you're **Catholic**, you'll have to confess — if you're **Chinese**, an hour later you'll have a nagging hunger to call back. But if you're **Polish**, go ahead — you probably dialed a wrong number.

BEEP...

HIGHEST PRIORITY

[RELIGIOUS MUSIC IN B.G.]

(DEEP VOICE)

Heaven... God speaking. You know, whenever I'm not busy answering prayers — I'm answering machines — that's a little joke.

Anyway, when you hear the tone, you better not hang up — I know who you are.

BEEP...

JEWISH MACHINE (FEMALE)

(JEWISH DIALECT - FAST)

This is <u>NAME's</u> machine. He's not home, already. Oy-vay, you should only have my headaches! You don't know from trouble! <u>NAME</u>, such a nice beautiful boy, he doesn't take care of himself, he eats like a bird. The poor boy's gonna get sick! And I'm not kidding when I tell you these strange women call at all hours, I don't know who they are. Oy — what's a poor Yenta machine to do??? Anyway, leave a message at the tone, but make sure it's kosher.

BEEP...

JEWISH MACHINE (MALE)

(JEWISH DIALECT - FAST)

Please leave a message and don't be a meschuggannah! This machine is a big investment—I spent a lot of hard earned money on it—so don't leave me hanging like my good-for-nothing cousin left me hanging — but that's another story — you don't wanna hear about it... Oy! But I'll tell you anyway...

[PAUSE]

What's this? A warning light? The tape's about to run out! Now I can't finish my story! Some big deal machine this is! No wonder everyone's hanging up — such a putz machine!

BEEP...

NEW-AGE MESSAGE

[PLAY MEDITATION MUSIC IN B.G.]

("SPACEY" VOICE)

Listen, I don't want to bore you with metaphysics, but how do you know this is really an answering machine? I mean maybe it's a dream — or maybe you don't even exist and this is all an illusion??? One way to find out: leave a message — if it's reality, I'll call you back.

BEEP...

PHONE POOLING

Hello. This is <u>NAME</u>. Because of my concern about our city's crowded phone lines, this week I'm answering machine pooling with Father O'Riley. So at the tone, leave a message — or a confession.

BEEP...

SALVATIONAL MESSAGE

(TV EVANGELIST VOICE)

Greetings, fren's. This is Brother <u>NAME</u> of **The Church Of The Perpetual Dialtone.** For lo', it is written in the Good Book, upon the Pages which are Yellow, that ye who leaveth a message shall have everlasting Call Waiting. And ye who hangeth up shall dwell in the underworld of Busy Signals for all eternity. So, dear fren', secure your salvation by leaving a message after the Sacred Tone. A-men!

BEEP...

STRENGTH OF PRAYER

Hello. You've reached "Dial-A-Prayer." Leave your name and number and **pray** I call you back.

BEEP...

THEOSOPHY

Hello. This is <u>NAME</u>. Something's been bothering me and I had to go have a serious talk with my pastor. Sunday, in church, he told the congregation that we were all put on Earth to help others. But he forgot to mention, why are the **others** here???

BEEP...

BUSINESS BUFFOONERY

(Dictation Was Never So Much Fun!)

AIRLINE

(STEWARDESS VOICE — FOR BEST EFFECT, HOLD NOSE)

Hello and welcome to <u>NAME</u> Airlines. We can't come to the phone right now because the Captain has turned on the FASTEN SEAT BELT sign. Please leave your name and number in its full upright position by grasping the phone receiver and placing it firmly over your mouth and nose. Once we reach our destination, we'll be sure to call you back. Please wait for the beep and assume the crash position.

BEEP...

CAR PHONEY

You have reached <u>NAME's</u> car phone... I know what you're thinking: just because he drives a little, bitty car, you don't believe there's a phone in it. Well, just to prove it, I will honk the little, bitty horn on his little, bitty car. Listen closely...

BEEP...

I DO SO HAVE A CAR PHONE!

COMPUTER CHALLENGE

(COMPUTER VOICE: "HAL" FROM "2001 - A SPACE ODYSSEY")

Hello. My name is Hal. I am a 2001 Model Computerized Answering Machine. I am a mega-billion times more intelligent than you humans. I am here to make you feel inferior and hang up. But if I have failed in my mission, leave your message at the beep.

BEEP...

DON'T LEAVE HOME WITHOUT...

Hi. Do you know me? Did I leave home without you? Be **American** — **express** yourself: leave your name, number and a **statement,** and I'll get back to you before your **expiration** date!

BEEP...

NEGATIVE MACHINE

(WHINY VOICE)

Hello. You've reached the Negativity Society. You can leave a message if you want, but it probably won't do you any good. Either we'll accidentally erase it, or we'll lose your number, or else we'll just forget to call you back. Actually, I don't know why we even bother to turn this stupid thing on???

BEEP...

NEWS REPORT

{PROP: TYPEWRITER}

[SINGLE TYPEWRITER KEY CLICKING IN B.G.]

(TV NEWS REPORTER VOICE)

NAME here. And this is your Newswire Answering Machine. **Flash:** late last night, thieves broke into the local police station and stole the toilet — the cops have nothing to go on!

[PAUSE]

And now, this just in from our Roving Reporter:

BEEP...

OBSCENE ETIQUETTE

(BRITISH ACCENT - "STUFFY")

It has come to our attention that many callers have not been apprised of the proper obscene-phone-call etiquette:

#1: Don't talk with your mouth full of saliva

#2: If you are calling from a pay phone, don't drool onto the mouthpiece— the next pervert may get your germs.

#3: Proper attire is essential — so be sure you are naked.

And finally: always leave a name and number so the slandered party can R.S.V.P.

BEEP...

PROCRASTINATION IS OUR BUS...

Hello. This is <u>NAME</u>, President of the Procrastinator's Society. At the tone, please leave your...

Wait a minute??? Isn't this is the same message I used yesterday???

Oh well, I'll change it tomorrow...

Nah, maybe next week.

BEEP...

(un)REAL ESTATE MESSAGE

(FRETTING)

Hello. I'm home but I can't pick up the phone. I'm so upset — I don't know what I'm going to do. I owe the bank a ton of money, I've lost almost all of my property and it looks like I may have to GO TO JAIL! Well, leave a message and I'll call you just as soon as we finish this...

Monopoly game.

BEEP...

REDUNDANT MESSAGE

Hello. You have reached the Assistant Secretary to the Vice-President of the Society of Repeating Things Over And Over Again Redundantly A Lot. Please be nice and kindly leave the name you're called and the digital integers of your phone number when you hear the sound of the beep tone noise. So long! Adios! Au revoir! Guten aben! Toodle-loo! See ya! Ta-ta! Bye-bye!

BEEP...

SALESMANSHIP

(EXCITED, OBNOXIOUS STEREO STORE SPOKESPER-
SON'S VOICE)

YOUR CITY, this is it! The home of **NAME**! You're listening
to his new **BRAND NAME** answering machine — **WITH**
beeperless remote. **NAME** has a **HUGE** inventory of errands
today — however, for a limited time, you can leave a
message — up to **30 SECONDS! UNBELIEVABLE!** Do it
NOW!

BEEP...

STOCK MACHINE

{PROP: TYPEWRITER}

[SINGLE TYPEWRITER KEY TICKING IN B.G.]

(FAST)

Good day. This is **NAME** with the Dow Jones Answering
Machine Report: Calls rose today in active phoning with
dial tones closing lower. Hang ups are declining rapidly
with messages rising sharply. Our Blue Chip barometer
shows return calls making a strong comeback in anticipation
of a bullish phone economy. So put down a short message
and we predict a big return call on your modest investment.

BEEP...

TEST LAB

<u>NAME</u>'s Laxative Testing Laboratory — <u>NAME</u> speaking. Right now I'm busy working my butt off. Leave a message, but don't give me any crap. I'll return your call, flush on the hour — if I'm not too wiped out. but believe me, I'm pooped already!

BEEP...

XEROXED MESSAGE

(CONCERNED)

Hi. Uh, listen, I had an electrical problem today and my answering machine got it's wires crossed with my copying machine...

So I hope you're getting this message...

    [PAUSE - SAME TONE OF VOICE]

So I hope you're getting this message...

    [PAUSE - SAME TONE OF VOICE]

So I hope you're getting this message...

BEEP...

HOLIDAY CHEER
(A Festive Fun-For-All)

January ... NEW YEARS - RESOLVE

Hello. It's New Year's so I made my resolutions and I'm kinda proud to say that last week I quit smoking, drinking and having sex — and let me tell you, it was the most horrifying **five minutes** of my life!

BEEP...

February ... LINCOLN'S BIRTHDAY - BUSH LEAGUE

Hi. It's Abe's birthday and I was just thinking how much former President Bush reminds me of Lincoln:

you see, both were Republicans; both had unattractive wives; and Bush's last score was seven years ago...

BEEP...

July ... INDEPENDENCE DAY - BLOW OUT

Hi! It's the 4th of July and I wanted to start the day off with a bang — unfortunately, my wife had a headache!

BEEP...

October ... COLUMBUS DAY - SAIL AWAY

Hi! It's Columbus day and I went sailing. By the way, almost everybody knows that Columbus set sail with three ships: the Nina, Pinta and Santa Maria. But few people know that only **two** ships made it back. On the return voyage, the Santa Maria **rear-ended** the Pinta and it blew up!

BEEP...

October ... HALLOWEEN - A HAUNTING EXPERIENCE

Hi. It's Halloween and I'm a little depressed. I'm thinking, maybe it's time for a facelift. This morning I went for a walk and everybody I met gave me candy.

BEEP...

November ... THANKSGIVING - GETTING THE BIRD

Hello. It's Thanksgiving, so don't be a turkey — leave your name and number.

BEEP...

November ... THANKSGIVING - QUIZ

{PROP: ENVELOPE}

This is <u>NAME</u> with your chance to be "Tarmac The Magnificent!"

I hold in my hand the envelope — and the answer is...

"The Thanksgiving Bird, The country next to Greece, and YOU if you hang up!"

[TEAR OPEN ENVELOPE]

And the question is...

BEEP...

December ... PRE-CHRISTMAS - BIG DISCOUNTS

Hello. This is <u>NAME</u> the kleptomaniac, just reminding you there's only (<u>SEVEN</u>) **shoplifting** days left until Christmas.

BEEP...

December ... CHRISTMAS - WRAP

Hello. This is <u>NAME</u>. I can't come to the phone because... I'm wrapping **your** very expensive Christmas gift. Please leave your name and number so I'll know **whose name** to put on the box!

BEEP...

December ... CHRISTMAS - A CAROL

'Twas the night before Christmas,
     and all through the house,
Not a creature was stirring,
     not even a mouse.
My answering machine was decorated,
     with a festive air,
In hopes that a message,
     soon would be there.
When all of a sudden,
     there rose such a clatter,
But it was only a hang-up,
     I cried "What's the matter?"
If you leave a message,
     when you call tonight,
It'll be a Merry Christmas for all,
     and for all a good night!

BEEP...

December ... CHRISTMAS - MALLED

Sorry ya missed me, I'm off to do some shopping — hadda get an early start. Maybe it's just me, but isn't it lame that they have Christmas during December when the malls are always so crowded???

BEEP...

... BIRTHDAY MESSAGE:

Hi. It's my birthday. That's right, I'm a year older and a lot slower — which means I probably won't get to the phone in time. So slide me a break and leave a message.

BEEP...

... SAFETY WEEK

Hi. It's National Answering Machine Safety Week, and I want to remind you: if you drink, don't leave a message! Have a friend leave it for you. Likewise, if you're calling from a party, have a designated speaker. Now, at the tone, talk carefully and have a safe phone call.

BEEP...

SPECIAL SITUATIONS
(Different Means For Different Machines)

CAT OWNER

(DEEP VOICE, SERIOUS)

Notice: the 110 volt current that runs this machine is wired to an adorable little kitten. Hanging up without leaving a message will complete the circuit and **fry the kitty!** It's your decision...

(IN BACKGROUND)

"Meow... meow... meow..."

BEEP...

DOG OWNER #1

(GOOFY VOICE)

Hello, This is NAME's dog. NAME is not home right now, so leave a message. And do me a favor — tell 'm I'm leavin' — for good! He's always goin' out — never leaves food or water! And I'm tired of fetchin' his smelly slippers. But before I go, I'm gonna leave him a little somethin' behind the couch to remember me by.

BEEP...

DOG OWNER #2

(TOUGH VOICE)

Hi. This is <u>NAME</u>. I'm gonna be gone for a while, so please leave your name and number. Oh, by the way, if there's any criminals listening, don't try to rob my apartment because I have a **really** vicious dog in here, right Spike?

(TOY POODLE)

"Ruff, ruff-ruff, ruff, ruff..."

BEEP...

EXPECTING UNWANTED CALL

Hey! I was expecting your call! That's why I left my machine on. That's also why I'm not home.

BEEP...

FINALLY, A NEW MESSAGE

{PROP: SNARE DRUM}

[DRUM ROLL IN B.G.]

And now... the moment you've all been waiting for...

[INCREASE VOLUME OF DRUM ROLL]

<u>NAME</u> has a new message!

BEEP...

HOME, BUT NOT ANSWERING THE PHONE #1

Hi. This is NAME. I'd really like to chat, but I had to step out of the room for a second. If I'm not back by the time you hear the tone, just start the conversation without me
....
....
....
BEEP...

HOME, BUT NOT ANSWERING THE PHONE #2

Hi. I'm home but I can't pick up the phone. I have to study all night. I wanna be sure and get an "A!" I got a big urine test, tomorrow!

BEEP...

HOME, BUT NOT ANSWERING THE PHONE #3

[POPULAR TV SHOW IN B.G.]

Please leave a message. I'm home, but due to technical difficulties, my feet are unable to get my mouth to the phone.

BEEP...

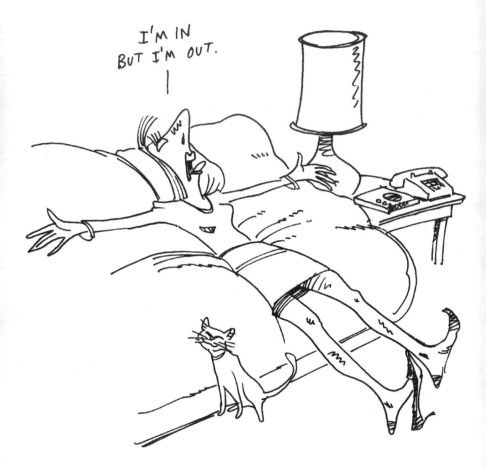

LIMITED LENGTH OF MESSAGE

Hello. You've reached "Dial & Moan." At the tone, leave your name, number and your **problems...**

Fortunately for me, you only have 20 seconds!

BEEP...

LIMITED LENGTH OF MESSAGE

(BUSINESS-LIKE)

Hello this is <u>NAME</u>. You have 20 seconds to leave your name, number and a brief message. But, if it's a dirty phone call...

(FLIPPANT)

talk as **long** as you want.

BEEP...

SMOKER'S SECTION

(BAD SMOKER'S COUGH)

Hi. At the tone, leave a message.

(COUGH-COUGH)

Excuse me! Damn! I gotta do something — and quick!

(COUGH-COUGH)

Every time I pick up a paper I **read** about the horrible side-effects of smoking,

(COUGH-COUGH)

so I've made up my mind — I'm gonna quit...

(COUGH-COUGH)

reading!

BEEP...

[NOTE: None of the other messages would consent to share this page.]

LONG ABSENCE

Hi. This is <u>NAME</u>. Please leave a message. I have to go to the **Post Office,** my **bank** and then the **DMV,** so you can expect a return call in about...

two weeks.

BEEP...

UNLIMITED TAPE

Hello. You've reached a **Van Gogh** Answering Machine. This machine has an unlimited tape, so if you want, you can talk my **ear** off!

BEEP...

VACATION MESSAGE #1

Hi. You've reached the <u>NAME</u>'s. We're gonna be out of town for a few days. You know, we've found that getting away every once in a while really helps our relationship. So **I** went to Vegas and she went to New York.

BEEP...

VACATION MESSAGE #2

(UPPER CLASS SNOBBISH VOICE)

-OR-

(HILLBILLY VOICE)

Sorry you missed us, but we're off on another European vacation — first class, of course. Last month we went to Venice, Italy — what a rip-off! It cost us over ten grand and when we got there, the whole place was flooded!

BEEP...

RANDOM FUN
(Hi-jinks For Every-One)

ATTITUDE

Yeah, yeah, I know, you've reached a lousy machine! But at least it's better than a maid who can't speak English!

BEEP...

"BAD" MEDICINE

[JUMP UP AND DOWN THROUGHOUT]

Hi. This is <u>NAME</u>. I hope you can understand me. I have to keep jumping up and down. I just took some real important medicine and forgot to read the label — it said, "Shake Well Before Using!"

BEEP...

CATEGORIZED CALLER

Studies indicate that people who hang up on answering machines are insecure, paranoid and usually psychotic. While people who talk to answering machines are secure, intelligent and successful. Please categorize yourself after the tone.

BEEP...

CHILDISH MACHINE

(THREE-YEAR-OLD VOICE)

Hell-wo. I'm Tommy Thumbnail and I'm a half-inch taw-ll. I wiv' inside dis machine an' take <u>NAME</u>'s caw-lls. Say, would-ja do me a favor, pwease? Weave a message, pwease? It's dark in here an' when ya weave a message, da widdle light comes on. Tanks, pal!

BEEP...

COIN-OPERATED MACHINE

{PROPS: TIN CUP or CAN — 3 QUARTERS}

[DROP QUARTER INTO CUP]

Hi. This is NAME.., Hang on a second...

[DROP QUARTER INTO CUP]

I have one of those old coin-operated answering machines. Damn thing is costing me a fortune!

[DROP QUARTER INTO CUP]

Anyway, at the tone, please leave..

(PINCH NOSE)

"This is the Operator. Please deposit another 75 cents if you wish to continue your message!"

(NORMAL VOICE)

Forget it Operator! I'm not about to pay another cent! And if you don't like it, you can shove this coin right up your...

BEEP...

CONSTIPATED MESSAGE

[RECORD IN OR NEAR BATHROOM]

(FAR-AWAY VOICE)

Hello. <u>NAME</u> here. If you leave your name and number, I'll get back to you very soon. I **am** home, but I can't come to the phone right now because... well, uh, I won't bore you with the details... let's just say I'm trying to work something out...

[FLUSH TOILET]

BEEP...

DENTAL MESSAGE

[ECHO EFFECT: TALK INTO A BOWL OR A POT]

HELLO... hello

This is <u>NAME</u>... <u>name</u>

I had to go to the DENTIST... dentist

I think I have a big CAVITY... cavity

BEEP...

GERMAN MACHINE

(GERMAN ACCENT)

You haf reached a **Cherman** answering machine, und you **vill** leaf a message! Ve haf ways of making you talk!

BEEP...

HICCUPS

Hi. (HICCUP) This is <u>NAME</u>. (HICCUP) Would you do me a (HICCUP) favor? At the (HICCUP) tone, please leave a **scary** message! (HICCUP)

BEEP...

HYPOCHONDRIAC MACHINE

(FAST AND NERVOUS)

Hi. I'm a Hypochondriac Model Answering Machine. Leave a message, but do it quick! I gotta go back in the shop again! I've had this nagging pain in my transistors all day — and yesterday my circuits were throbbing like you wouldn't believe! Plus, I'm all out of my special lubrication — not to mention my weak resistors! Oh, and I almost forgot to tell you about my reoccurring...

BEEP...

IDIOTIC MESSAGE

This is the electronic idiot taking your call. You can give the electronic idiot a message for someone. You can have the electronic idiot ask someone to call you back. Or, you can call again later in hopes that a real live idiot will answer.

BEEP...

IDLE THREAT

(ANGRY)

This is <u>NAME</u>. There's been a lot of hang-ups lately and I'm really getting fed up! I don't know who you are, but if you don't leave your name and number I'll... I'll... Uh??? I'll never speak to you again!

BEEP...

MEDICINAL MESSAGE

Hi. You say answering machines give you a headache? Well, leave two messages, get plenty of sleep and I'll call you in the morning.

BEEP...

MENTALLY CHALLENGED MESSAGE #1

Hello. this is <u>NAME</u>. You know, I make a lot of crude jokes, but I never make fun of the handicapped — because if it wasn't for them, I'd never get good parking spaces.

BEEP...

MENTALLY CHALLENGED MESSAGE #2

Hello. This is <u>NAME</u>. Leave a message at the tone and I'll get back to you — for those who don't have lips:

'eave a 'essage an' i 'all 'ou 'ack. 'ank 'ou!

BEEP...

PERSONAL AD

(MONOTONE)

Help Wanted:

Responsible person to leave important message. Must be intelligent, able to recite their seven digit number and pronounce own name. Inquire within.

BEEP...

...OR BEST OFFER

Hi. I ran out to the video store! Leave a message. The other night I saw this movie on video called "Indecent Proposal" where Demi Moore goes to bed with a stranger for a million bucks. Of course, I can't compete with Redford, but I'm curious, at the tone, tell me: if I slip you a **ten-spot** what would **you** do?

BEEP...

PHONE POLICE

(COP VOICE - DEEP & SERIOUS)

This is Officer <u>NAME</u>. You have the right to remain silent. If you speak, the message you leave may be used to call you back. If you do not have a message, the court will appoint you one.

BEEP...

PLUMBING PROBLEMS

[FOR BEST RESULTS, RECORD IN OR NEAR YOUR BATHROOM]

(CONCERNED)

Hello. This is <u>NAME</u>.

[RUN WATER]

Hope you get this message.

[BANG ON PIPES]

My machine broke last week and I made the mistake of letting my brother-in-law, the **plumber,** fix it.

[FLUSH TOILET]

BEEP...

PSYCHIC MESSAGE

Hello. Do you believe in psychic powers? 'Cuz I know exactly why you called. The reason is... I just stepped into the shower!

BEEP...

SCHIZO MESSAGE

(FIRST VOICE)

Hello. This is NAME.

(SECOND VOICE)

Hello. This is the **other** NAME.

(FIRST VOICE)

Listen, I know that lately a lot of my friends have been a little worried about my...

(SECOND VOICE)

mental health.

(FIRST VOICE)

But I look at it this way: I may be schizophrenic...

(SECOND VOICE)

but at least I'm never alone.

BEEP...

SELF ANALYSIS

(PROFESSIONAL ANNOUNCER VOICE)

Good day. Scientists have recently made a direct connection between the fear of talking to a machine and sexual dysfunction. If you hang up, I won't know who you are, but **you'll** know what your problem is.

BEEP...

SINGING MESSAGE

You know, there must be 50 ways to leave a message...

Just call me back, Jack—

Tell me your plan, Stan—

You don't need a ploy, Roy—

Just call on me.

Give me a buzz, Gus—

You don't have to discuss much—

Just say something brief, Leif—

And you'll be hearin' from me!

BEEP...

SESQUIPEDALIAN (look it up) MESSAGE

(INTELLECTUAL VOICE)

I'm not home right now, so leave a message. However, when promulgating your esoteric cogitations, beware of the platitudinous use of loquacious vernacular!

(NORMAL VOICE)

Uh, I'm sorry... just don't use any big words, okay?

BEEP...

SMART MACHINE

(CURT TONE)

Hi. This is <u>NAME</u>. Leave a message. I'm outta here.

(NEW VOICE - CONSPIRATORIAL WHISPER)

Hi. This is <u>NAME's</u> machine. Now that he's gone we can talk about him! Boy, you know what I hate? First thing every morning when he speaks into my mike — talk about bad breath! So... tell me what he does that bugs you... and don't worry, it's between you and me — I won't tell!

BEEP...

WET MESSAGE

[SPEAK WITH WATER IN YOUR MOUTH, MAKING A GARGLING SOUND]

Hi. This is <u>NAME</u>. Please help me out and leave a message. I'm testing out my new water-proof machine!

BEEP...

ZE LAST MESSAGE IN THE BOOK

(SLOW, SLURRED AND LOW-TONED)

H e l l o o o o o o o o. T h i s is <u>NAME</u>. I had to go to the store and buy s o m e new batteries f o r t h i s s s s s s s s s s s s

m a c h i n n n n n n n n n e...

BEEP...

TITLES BY CCC PUBLICATIONS

RETAIL $4.99

CAN SEX IMPROVE YOUR GOLF?
THE COMPLETE BOOGER BOOK
THINGS YOU CAN DO WITH A USELESS MAN
FLYING FUNNIES
MARITAL BLISS & OTHER OXYMORONS
THE VERY VERY SEXY ADULT DOT-TO-DOT BOOK
THE DEFINITIVE FART BOOK
THE COMPLETE WIMP'S GUIDE TO SEX
THE CAT OWNER'S SHAPE UP MANUAL
PMS CRAZED: TOUCH ME AND I'LL KILL YOU!
RETIRED: LET THE GAMES BEGIN
MALE BASHING: WOMEN'S FAVORITE PASTIME
THE OFFICE FROM HELL
FOOD & SEX
FITNESS FANATICS
YOUNGER MEN ARE BETTER THAN RETIN-A
BUT OSSIFER, IT'S NOT MY FAULT

RETAIL $4.95

1001 WAYS TO PROCRASTINATE
THE WORLD'S GREATEST PUT-DOWN LINES
HORMONES FROM HELL II
SHARING THE ROAD WITH IDIOTS
THE GREATEST ANSWERING MACHINE MESSAGES OF ALL TIME
WHAT DO WE DO NOW?? (A Guide For New Parents)
HOW TO TALK YOUR WAY OUT OF A TRAFFIC TICKET
THE BOTTOM HALF (How To Spot Incompetent Professionals)
LIFE'S MOST EMBARRASSING MOMENTS
HOW TO ENTERTAIN PEOPLE YOU HATE
YOUR GUIDE TO CORPORATE SURVIVAL
THE SUPERIOR PERSON'S GUIDE TO EVERYDAY IRRITATIONS
GIFTING RIGHT

RETAIL $5.95

50 WAYS TO HUSTLE YOUR FRIENDS ($5.99)
HORMONES FROM HELL
HUSBANDS FROM HELL
KILLER BRAS & Other Hazards Of The 50's
IT'S BETTER TO BE OVER THE HILL THAN UNDER IT
HOW TO REALLY PARTY!!!
WORK SUCKS
THE PEOPLE WATCHER'S FIELD GUIDE
THE UNOFFICIAL WOMEN'S DIVORCE GUIDE
THE ABSOLUTE LAST CHANCE DIET BOOK
FOR MEN ONLY (How To Survive Marriage)
THE UGLY TRUTH ABOUT MEN
NEVER A DULL CARD
RED HOT MONOGAMY (In Just 60 Seconds A Day) ($6.95)

RETAIL $3.95

YOU KNOW YOU'RE AN OLD FART WHEN...
NO HANG-UPS
NO HANG-UPS II
NO HANG-UPS III
GETTING EVEN WITH THE ANSWERING MACHINE
HOW TO SUCCEED IN SINGLES BARS
HOW TO GET EVEN WITH YOUR EXES
TOTALLY OUTRAGEOUS BUMPER-SNICKERS ($2.95)

NO HANG-UPS – CASSETTES   RETAIL $4.98

Vol. I: GENERAL MESSAGES (Female)
Vol. I: GENERAL MESSAGES (Male)
Vol. II: BUSINESS MESSAGES (Female)
Vol. II: BUSINESS MESSAGES (Male)
Vol. III: 'R' RATED MESSAGES (Female)
Vol. III: 'R' RATED MESSAGES (Male)
Vol. IV: SOUND EFFECTS ONLY
Vol. V: CELEBRI-TEASE